Heat

Energy That Flows

Glen Phelan

Sally Ride, Ph.D., President and Chief Executive Officer;
Tam O'Shaughnessy, Chief Operating Officer and
Executive Vice President; Margaret King, Editor;
Monnee Tong, Design and Picture Editor; Erin Hunter,
Science Illustrator; Brenda Wilson, Editorial Consultant;
Matt McArdle, Editorial Researcher

Program Developer, Kate Boehm Jerome
Program Design, Steve Curtis Design Inc.
www.SCDchicago.com

Sally Ride Science is a trademark of Imaginary Lines, Inc.

Sally Ride Science
9191 Towne Centre Drive
Suite L101
San Diego, CA 92122

ISBN: 978-1-933798-54-7

Printed in the United States of America
10 9 8 7 6 5 4 3 2 1
First Edition

Cover: A hot-air balloon drifts up because the warm air
inside is less dense than the cooler air outside.

Title page: The soaring temperatures in Namibia's Dead
Valley mean the air molecules there have a lot of energy.

Right: Water boils when heat is conducted from the
burner to the pot and then to the water molecules.

Contents

Introduction

In Your World 4

Chapter 1   **Heat 101**

All About Motion 6

Chapter 2   **Temperature and Heat**

Going With the Flow 10

Chapter 3   **Heat Transfer**

Energy on the Move.................................16

Thinking Like a Scientist

Predicting 24

Interpreting Data 25

How Do We Know?

The Issue

A Hot Problem in Cold Space................26

The Expert

Jay Ochterbeck 27

In the Field...28

Technology..29

Study Guide

Hey, I Know That! 30

Glossary 31

Index 32

In Your World

These huge dinosaurs are the hit of the snow-sculpture competition. Before you know it, however, they'll be puddles! How come? While thinking about this, let's head inside and warm up in front of the fireplace before we see the other sculptures.

Ahh! Nothing beats a crackling fire on a bitter cold evening. The flames cast a warm glow on your skin. It feels good. So does the mug of hot soup you wrap your hands around. Each sip warms you from the inside out.

Don't rest too long! Today may be your last chance to check out the other snow sculptures. Tomorrow is supposed to be much warmer, and you know what that means. These works of art will melt away in hours. What makes them melt, and what actually happens when they do? How do the fire and hot soup warm you up? What's the real difference between hot and cold? There's more to the answers than you may think.

All About Motion

▶ *Mmm-mmm*, **good. This tasty treat packs a lot of heat because of motion you can't see.**

Do you know how popcorn pops? The hard kernels contain water. When the kernels get really hot, like they do in a microwave oven, the water inside turns to steam. The steam rushes out of the kernels, and they burst open into the fluffy white puffs you like to munch at the movies.

What makes freshly popped popcorn so hot? You may think the answer is simple—the oven. But the truth is more interesting than that. Motion makes the popcorn hot—not the motion *of* the popcorn, but the motion *inside* the popcorn.

Look around you. Everything is made of **molecules**—flowers, cars, popcorn—everything. You can't see the molecules, but if you could, you'd see that they're moving. In fact, an object's molecules are *always* moving, even when the object isn't.

This Way and That Way

What's moving in the picture below? A lot more than you think! The molecules that make up any kind of **matter** are constantly in motion. Molecules in a gas zip in all directions. Molecules in a liquid slip and slide past each other. Molecules in a solid are held in place, but they **vibrate** back and forth. It's like they're running in place—they move without going anywhere.

The molecules in unpopped popcorn are moving even before you start the microwave oven. But when the microwave kicks into gear, the molecules really start kicking!

In the oven, microwaves bombard the popcorn. Water in the kernels absorbs this **energy**. The added energy makes the water molecules move faster and faster. The faster they move, the warmer the water and the kernals get. Eventually the molecules move so fast that the water becomes a hot gas, which bursts from the popping corn.

So the microwave oven doesn't make the popcorn hot. Microwaves make the molecules in the kernels move faster, and *that* makes the popcorn hot.

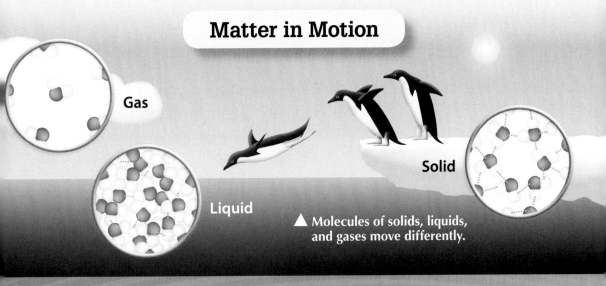

Matter in Motion

Gas

Liquid

Solid

▲ Molecules of solids, liquids, and gases move differently.

The Bottom Line

The faster the molecules in a solid, liquid, or gas move, the warmer the substance becomes.

Bringing the Heat

An object gets warmer when its molecules absorb energy and move faster. The object gains **heat**. Now there's a word you hear a lot—*heat*. What exactly does it mean?

The most important thing to remember about heat is that it's a form of energy. It's the energy in all the zipping, sliding, and jiggling molecules that make up an object.

Like all forms of energy, heat causes changes. It changes a cooler object into a warmer one. It cooks food, boils water, melts ice, and can even buckle roads on really hot days. In short, heat makes things happen.

Other forms of energy make things happen, too. And when they do, they usually produce heat. You're a great example! After you eat, your body uses the chemical energy stored in food to keep all your cells working like they should. In the process, your body produces the heat that keeps you nice and warm inside.

▲ The spinning and rubbing of the drill produces a lot of heat.

▲ In this special image taken with an infrared camera, red, orange, and yellow show the areas that give off the most heat. Green and blue show areas that give off less heat.

The biggest source of heat may be right over your head—the Sun! It provides energy for trees and other plants. Some trees are cut down for fuel. What happens when we burn the wood? It gives the energy back as heat!

How Much?

You've probably heard of **calories**. We use calories to measure the amount of food we eat, right? Well, actually, we use calories to measure the amount of *energy* the food contains. The calorie is a unit for measuring energy. Since heat is a form of energy, heat can also be measured in calories.

The main unit for energy—and heat—is the **joule**. One calorie is equal to a little more than 4 joules. One joule of heat isn't much. You give off 100 joules of heat every second just by sitting still and reading this book. *Hmm . . .* how many calories is that?

The Bottom Line | Heat is a form of energy.

Chapter 2: Temperature and Heat

Going With the Flow

It's a very hot day. You and your friends are hanging out at the neighborhood pool. While wiping the sweat off your brow, you say, "Wow! It sure is hot today!" If you really want to impress your friends, you could add, "The average energy of the air molecules is off the charts."

Okay, it may not impress your friends, but it would impress your teacher. It would show that you really know what **temperature** means.

Look at the glass of iced tea on the next page. You can't see its molecules, but some are moving faster than others.

A speedy molecule in the tea might collide with a slower one. When it does, it loses some of its speed and makes the slowpoke move faster.

Molecules are always colliding and changing speeds. We can't measure the motion of each molecule, but we can measure the *average* motion of all of them. The measure of that average energy of motion is temperature!

Temperature's Rising

Which has a higher temperature— the hot tea or the iced tea? You don't need a science book to tell you it's the hot tea, but now you know why. The molecules are moving faster in the hot tea than in the iced tea, so they have a higher average energy of motion. That's what we mean when we say the hot tea has a higher temperature.

What happens if you heat the iced tea on the stove or leave it out in the Sun? The added energy makes the molecules move faster, and the tea heats up.

▲ **Which tea has the faster-moving molecules? Why?**

Molecules in Motion

This photo shows two glasses of water with food coloring added. Can you guess which glass contains hot water and which contains cool water? If you guessed that the glass with the blue food coloring holds hot water, give yourself a pat on the back. The faster-moving molecules of hot water collide harder with the molecules of food coloring and make them move faster. So the food coloring swirls and spreads out faster.

The Bottom Line

The temperature of an object is the measure of the average energy of motion of its molecules.

Why does ice make your hands feel cold?

On the Move

What happens when you hold an ice cube in your hand? It doesn't stay ice for long. Within seconds, water begins dripping through your fingers. The ice is cold, the water is cold, and your hand feels cold. Yet this is a perfect example of heat!

You already know that heat is a form of energy, but there's more to it than that. Heat is energy that flows from an object of higher temperature to an object of lower temperature. Your hand is at body temperature—about 37°C (98.6°F). That's a lot warmer than ice. So as soon as you touch the ice cube, energy flows as heat from your hand to the ice. How? It all comes back to the motion of molecules.

The speedy molecules that make up the skin of your warm hand smash into the slower molecules on the surface of the ice cube. This slows down the molecules in your hand and speeds up the ones in the ice. Your hand *loses* energy, so it feels colder. The ice *gains* energy, so it warms up and melts.

So That's Why!

"The ice made my hand cold." That's a natural thing to say, but it's not really accurate. The ice doesn't give your hand coldness. There's no such thing as "coldness." Your hand feels cold because heat moves away from it. So the ice doesn't make your hand cold—the loss of heat does.

Heat flows from hot food to the cooler matter around it.

Cool Off!

Have you ever been told to eat your food before it gets cold? Well, here's something that might surprise you. The food won't really get cold. Here's why.

Heat flows from warmer objects to cooler ones until both are the same temperature. Your bowl of hot soup cools off as heat moves from the soup to the cooler bowl and air. The food cools off, while the bowl and air warm up. The heat flows until everything is room temperature. So the soup won't get colder than the temperature of the room.

But why does left-out food *feel* colder than the air in the room? It's because your body is warmer than the food. If you stick your finger in the soup after it sits out for a couple hours, what happens? Heat flows from your finger into the soup. Because heat leaves your finger, your finger feels cool.

The Bottom Line | Heat flows from warmer objects to cooler ones, never the other way.

Temperature Check

How would you measure the temperature of water in a science experiment? With a **thermometer**, of course!

One kind of thermometer is made of alcohol with red coloring sealed inside a narrow glass tube. When the thermometer touches the water, the water molecules collide with the molecules in the glass walls of the thermometer. Suppose the water is warmer than the thermometer. Then the water molecules have more energy of motion than the molecules that make up the thermometer. As the molecules collide, the ones in the water pass some of their energy to the ones in the thermometer. These molecules gain energy and speed up.

In this way, energy flows as heat from the water to the glass tube and then to the alcohol inside. The alcohol molecules speed up and spread out, and the alcohol expands. The alcohol can go only one place—up the tube! The alcohol will continue to heat up until—but only until—it's the same temperature as the water.

▼ **This scientist uses a thermometer to measure a stream's temperature.**

What if the water is cooler than the thermometer? Then the molecules in the thermometer pass some of their energy to the water. The alcohol molecules slow down and move closer together. The alcohol level drops until the alcohol is the same temperature as the water.

Choose Your Scale

The alcohol in the thermometer rises or falls to a certain point. What does that tell you? Not much, until you compare the level to the temperature scale.

The two main temperature scales are degrees **Celsius** (°C) and degrees **Fahrenheit** (°F). The Celsius scale is based on the temperature at which water freezes and boils. On the Celsius scale, 0 is the freezing point of water, and 100 is its boiling point. So every degree Celsius is 1/100th of the difference between the freezing point and boiling point of water.

On the Fahrenheit scale, water freezes at 32 degrees and boils at 212 degrees. Compare other temperatures from the two scales shown on the thermometer.

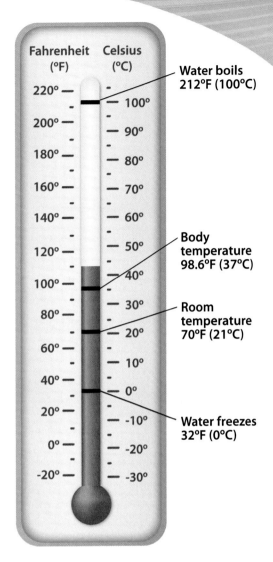

Water boils 212°F (100°C)

Body temperature 98.6°F (37°C)

Room temperature 70°F (21°C)

Water freezes 32°F (0°C)

▲ Just like centimeters and inches are different units for measuring length, degrees Celsius and degrees Fahrenheit are different units for measuring temperature.

The Bottom Line

Thermometers work because energy moves as heat from a warmer object to a cooler object until both are the same temperature.

Energy on the Move

Here's a really cool trick. Put an empty, uncapped glass bottle in the freezer until it is nice and cold—an hour should do. Then take the bottle out and set it on a table. Wet one side of a coin and place it, wet-side down, over the opening of the cold bottle. This will make an airtight seal. Now watch the coin carefully. Within a minute, one side of the coin will pop up once and then again and again. It's like the coin has the hiccups! Call it the hiccuping coin trick. Can you figure out how this trick works? The air in the room is warmer than the cold bottle. As soon as you take the bottle out of the freezer, the speedy molecules of warm air bombard the bottle. First the collisions make the molecules in the glass vibrate faster. Then those vibrating molecules collide with the air molecules inside the bottle, and these also begin to move faster.

The air in the bottle heats up. As the air molecules inside the bottle pick up speed and spread farther apart, they bang against the coin harder. Eventually, they hit it hard enough to pop it up—the hiccuping coin!

◀ **Be sure the glass bottle you use for the hiccuping coin trick has a narrow neck.**

Reach Out and Touch

The hiccuping coin trick shows that heat can move by **conduction**. *Huh?* You thought heat moves because molecules collide. Well, that's what conduction is. When warmer matter touches cooler matter, the molecules where they touch collide. The transfer of heat that occurs is called conduction.

You know some examples of conduction already. Remember the alcohol thermometer? It works when the molecules in the glass tube touch those in the water. Did you ever warm your hands by wrapping them around a steaming mug of hot chocolate? You were warming them by conduction. Food on the stove cooks by conduction, too. Heat passes from the flame or hot coil to the pan to the food. What other examples come to mind? In each example, think about what happens to the molecules.

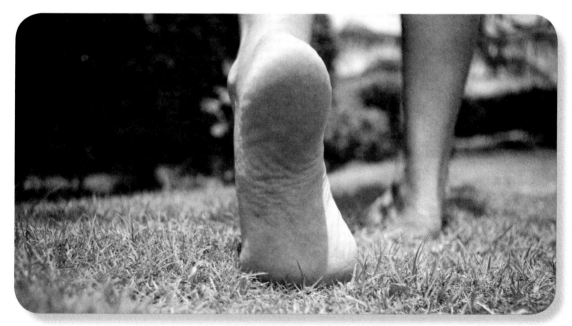

▲ Heat moves by conduction from a person's feet to the grass. So the grass feels cool.

The Bottom Line	When two things touch, heat moves from the warmer one to the cooler one by conduction.

▲ *Ouch!* **Don't even think about grabbing these pot handles with your bare hands. Why?**

A Quick Transfer

Mmm . . . the chili smells terrific! It's time to give it a little stir. A metal spoon is leaning against the side of the pot, but you'd better not grab it with your bare hand. Why? Metal is a **conductor**.

Some materials transfer, or conduct, heat better than others. Metals, like steel and copper, are the best conductors. A hot pot of chili shows that! Heat moves quickly from the metal pot to the metal spoon touching it.

You can tell that metal is a good conductor not only by how hot it feels but by how cold it feels. Look around right now for something that's metal. A doorknob is a good choice. Go ahead and touch it. It feels cool, doesn't it? It probably feels cooler than the wooden door, your pencil, or this book. Yet all these things are the same temperature— the temperature of the room. The doorknob feels cooler because the metal quickly carries heat away from your hand when you touch it.

Nice and Slow

Why is a wooden spoon a better choice for stirring the chili? The answer is that it doesn't transfer heat easily. Things like this are called **insulators**.

Wood, cloth, paper, and plastic are good insulators. So is air! In fact, air is what makes plastic foam cups and coolers work so well. Plastic foam is loaded with air pockets. The air traps heat and keeps it from flowing into or out of the cup. So hot things stay hot, and cold things stay cold.

So That's Why!

How can polar bears and walruses survive in the extreme cold? They have natural insulators. Fat and fur keep them warm by preventing body heat from escaping. Polar bears have a supersecret weapon. Their fur hairs have hollow spaces filled with air to give them extra insulation!

◀ Birds take advantage of air, too. This cardinal is fluffing out its feathers to trap a little air— and to trap body heat.

The Bottom Line | Conductors, such as metals, transfer heat well, while insulators, such as wood and air, do not transfer heat well.

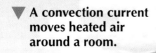

▼ **A convection current moves heated air around a room.**

Cool air

Furnace

Warm air

What Goes Around, Comes Around

Suppose you grab a baked potato fresh from the oven. Careful—that's one hot potato! The heat flows to your hand by conduction. But the potato molecules don't leave the potato and move into your hand. Conduction transfers energy even though the matter doesn't actually move from one place to another.

In liquids and gases, however, heat can also flow in another way. Matter can actually move from one place to another and carry its heat with it. This kind of heat flow is called **convection**.

Convection probably heats your home and your school. Warm air from a furnace enters a room through a vent or along a baseboard. The warm air molecules move fast and spread out, making the warm air less dense than the cooler air nearby. The warm air rises. Then it cools and sinks. This motion sets up a circular flow of air.

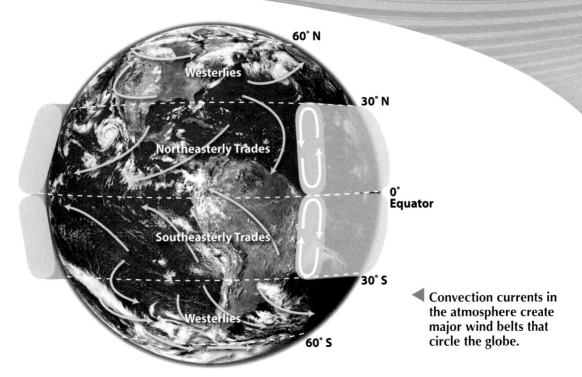

60° N

Westerlies

30° N

Northeasterly Trades

0°
Equator

Southeasterly Trades

30° S

Westerlies

60° S

◀ Convection currents in the atmosphere create major wind belts that circle the globe.

Mega Currents

Convection does a lot more than spread warm air around a room. It spreads warm air around our entire planet!

What gets these mega currents going? Is it a giant furnace blasting heat somewhere? Well, sort of. It's the Sun! The Sun heats all of Earth's surface, but the area around the **equator** gets the most direct sunlight. So this part of the planet gets the hottest.

Conduction transfers the heat from the ground to the air that touches it.

Then, just like in a room, this warm, less dense air rises, and cooler, denser air rushes in. This sets up humongous currents. The warm air travels thousands of kilometers before it cools enough to sink. So, if you live in the U.S., thank convection for moving some of the Sun's heat your way.

Convection spreads hot air and cool air all around the world. These flowing rivers of air create winds, bring the weather, and generally make our planet a pretty nice place to live.

The Bottom Line | Currents of warm gas or liquid carry heat from one place to another in a process called convection.

A Long Journey

Think about sunshine warming your face. How does this heat reach you from millions of kilometers away? You certainly didn't touch the Sun, so conduction is out of the question. A gas or liquid didn't carry the heat here, so it's not convection. The energy that warmed your skin was carried all the way from the Sun by **radiation**.

Radiation is the transfer of heat through waves of energy, such as visible light, ultraviolet light, or microwaves. A hot object, like the Sun, can transfer heat to a cooler object, like Earth, even though that object is some distance away.

Radiated energy, like light waves, can travel through empty space. That's a good thing. This kind of energy transfer provides the sunlight we need in order to see. Radiation provides the energy that heats the ground, which then heats the air. Energy from radiation even makes plants grow. In fact, most of the light and heat you experience come from the Sun's radiation.

▼ **The Sun radiates its energy out into space. These flowers intercept some of that energy.**

▲ Heat flows from a fire by radiation.

When you zap a plate of broccoli in a microwave oven, why doesn't the plate get as hot as the broccoli? It's all about radiation. The oven radiates microwaves. They make water molecules in the broccoli move like crazy. That heats up the broccoli. But the plate doesn't contain water, so it doesn't get as hot.

Radiation—Not so Scary

Most radiation that we experience is harmless. After all, where would we be without the Sun? Everything is constantly giving off energy in the form of radiation—the Sun, a fire, your desk, even you! The hotter the object is, the more energy it radiates.

Energy gets carried through space until it hits something that absorbs it. That might be a driveway absorbing radiation from the Sun, or your body absorbing radiation from a fire. Either way, energy transfers from a hotter object to a cooler one.

From the blazing Sun to a crackling fire to melting snow sculptures, heat is a part of your everyday life.

The Bottom Line | **Radiation carries energy through space.**

THINKING LIKE A SCIENTIST

***O**oh! Aah!* The hot sand feels like balls of fire under your feet. You run across the beach toward the water. As an incoming wave laps at your ankles, you suddenly stop. The water is too cold to go farther.

How can the sand be so hot and the water be so cold? If you've ever asked this question, or any question about things you observed, you're thinking like a scientist.

So what's the answer? It involves a material's ability to take in or give off heat. Some materials, like water, can take in a lot of heat without experiencing a big change in temperature. The Sun can beat down on a lake all day long, and its temperature won't change much.

Sand, on the other hand, does not require a lot of heat to change its temperature. It takes only a couple of hours in the Sun for sand to become scorching hot.

Predicting

Scientists have measured how much heat is needed to raise the temperature of 1 gram of different materials by 1 degree Celsius. This value is known as a substance's **specific heat**. The higher this number, the more heat is required to raise a material's temperature. In other words, the higher the number, the more slowly the material warms up. You can use these data to make predictions. Let's try it.

The specific heat for copper is 0.4 joules per gram degree Celsius. The specific heat of aluminum is 0.9 joules per gram degree Celsius. Do you think food would cook quicker in a copper pot or aluminum pot? Copper wins that contest. The lower specific heat means it heats up faster than the aluminum and passes the heat along to the food.

▼ **Which is warmer, the sand or the water? Why?**

Interpreting Data

Here are some other materials and their specific heat—how much heat is needed to raise the temperature of 1 gram of that material by 1 degree Celsius. It's a lot easier to compare these numbers and make predictions if you make a bar graph from these data.

How Much Heat?	
Material	Specific heat (joules/gram °C)
Aluminum	0.9
Copper	0.4
Glass	0.8
Iron	0.5
Plastic foam	1.3
Sand	0.8
Water	4.2

Your turn! Use your bar graph and the information on this page to answer these questions.

1. Which material needs the most heat to change its temperature? What is one effect of this?

2. Which three materials need about the same amount of heat to change their temperatures?

3. Which material needs the least heat to change its temperature? What is one effect of this?

4. A glass vase and a copper pot are on a fireplace hearth. They are the same distance from the fire. Predict which object will be warmer and explain why.

THE ISSUE

A Hot Problem in Cold Space

Have you ever balanced a laptop computer on your thighs? Then you probably felt the computer's toasty warmth. Switch on any electronic device and it will generate heat. Why? Electricity passing through the mazes of tiny wires on computer chips generates heat. That means the chips inside even the smallest laptop can quickly get hot to the touch. All electronic devices must get rid of that heat. If not, the heat can build up and damage delicate circuitry.

That makes heat a big problem . . . Enter a technology called a heat pipe. Heat pipes sound hot, but they cool down electronic devices on Earth—and in space. Heat pipes prevent meltdowns on space satellites that beam television signals to Earth. How? The electronics in a satellite warm one end of a sealed heat pipe through *conduction*. Inside, ammonia or another liquid boils and evaporates. The vapor carries heat to the other end of the pipe through *convection*. There, the vapor condenses, releasing the heat. The heat moves out of the heat pipe and into space through *radiation*. The liquid returns through the pipe with the help of a material that draws it back. The process is very clever—heat pipes use all three heat transfer methods!

▼ *Cool!* Thanks to the heat transfer methods in heat pipes like this, space satellites don't have meltdowns.

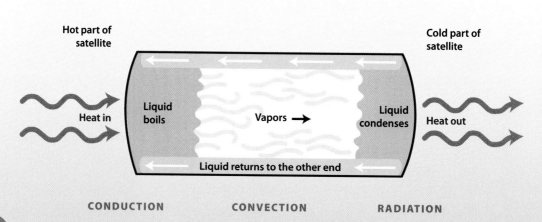

Hot part of satellite

Cold part of satellite

Heat in

Liquid boils

Vapors →

Liquid condenses

Heat out

← Liquid returns to the other end

CONDUCTION   CONVECTION   RADIATION

Jay Ochterbeck

CLEMSON UNIVERSITY

Mechanical Engineer

◀ Jay and his daughter enjoy a day at an amusement park.

Jay Ochterbeck grew up in New Mexico, where he learned to love the heat—of spicy barbecued food, that is! Jay also loved science and math. His favorite hobby was building model airplanes and ships. "I could always just visualize how things fit together," Jay says.

Jay's fascination with how things work led him to study mechanical engineering. Soon Jay was hooked on heat pipes. He was amazed at how quickly they could hustle heat from one end to another. "It was an area with lots of theories and math, but it could be applied to real problems, like keeping satellites cool in space," he says. Today Jay splits his time between teaching engineering classes and doing research on heat pipes—whether they're used to cool PlayStations or satellites.

For fun, Jay still heats up the barbecue to cook favorites like brisket of beef. "You can't cook it too quickly or too slowly," Jay explains, sounding like both a chef and an engineer. "It's all a heat transfer process."

▲ Jay shows a student a heat pipe that can cool the large batteries used in electric cars.

Jay Ochterbeck hands each laboratory team a mug of hot water and two cold metal rods. He tells them that one rod is solid copper and the other is a heat pipe with only a thin copper skin.

"Which one is which?" a student asks.

"That's what you're about to find out," says Jay. "Get stirring!"

Jay's students stir their steaming mugs, first with one rod and then the other.

They look amazed. One student checks the time. "One rod took 10 seconds to get hot—the other just 1 second!" she says.

"That means one rod transports heat 10 times faster than the other!" adds her lab partner.

Jay smiles. "Not even a good conductor like copper transfers heat as quickly and efficiently as a heat pipe!"

A large satellite such as this one might rely on more than 100 heat pipes to stay cool. "They're a great way to keep things cool in space," Jay says. "They have no pumps, no fans, and no moving parts—there's nothing to burn out and nothing to fix." That's one cool tool!

INVESTIGATION CONNECTION

Wet or Dry?

Would a heat pipe get rid of more heat if it was filled with air instead of liquid? Try this experiment in class. First everyone puts on safety goggles. Next a volunteer inflates a balloon and ties it to a chair. Then watch as your teacher holds a candle or Bunsen burner under the balloon. What happens? Predict what would happen if the balloon was filled with water. Then go outside and do the experiment with a water balloon.

Lots of Watts

Satellites orbit Earth in the freezing cold of space—so why is heat a problem? Satellites are real energy hogs, that's why! A big communications satellite packed with electronics might use as much as 15,000 watts of electricity. That's 1,000 times more watts than a laptop computer uses! That's a lot of heat to get rid of.

Hey, I Know That!

You've learned a lot about heat—what it is, how it moves, and how it affects your life. On a sheet of paper, show what you know as you complete the activities and answer these questions.

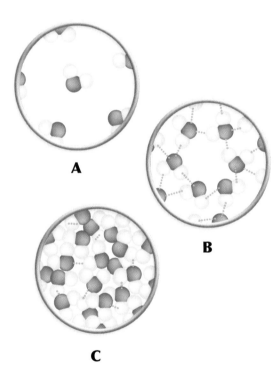

A

B

C

1. Look at the three pictures of molecules. Which picture shows a gas? How do you know? (page 7)

2. Which picture shows water after most of its heat has flowed away from it? Explain your answer. (page 7)

3. How does food help keep you warm? (page 8)

4. Suppose you are told it's 25 degrees outside. What else do you need to know in order to decide if you need to wear a coat? Why? (page 15)

5. Draw one or more pictures that show the three ways that heat is transferred. You may combine more than one kind of heat transfer on one picture. Include arrows and labels in your drawings. (pages 16–23)

6. Explain how your hand becomes warm when you stick it in a sink filled with warm water. Use the terms "molecules" and "energy of motion" in your answer. (page 17)

Glossary

calorie (n.) the amount of heat energy needed to raise the temperature of 1 gram of water 1 degree Celsius—also the amount of heat energy that 1 gram of water releases when it cools down by 1 degree Celsius. A Calorie is actually a kilocalorie and is used to indicate the energy content in food. (A kilocalorie is 1,000 calories.) (p. 9)

Celsius (adj.) referring to the temperature scale in which the freezing point of water is 0 degrees and the boiling point of water is 100 degrees (p. 15)

conduction (n.) the transfer of heat from one molecule to another (p. 17)

conductor (n.) any substance through which electricity or heat easily flows (p. 18)

convection (n.) the motion of a fluid due to changes in density from changes in temperature (p. 20)

energy (n.) the ability to cause change and make something happen (p. 7)

equator (n.) an imaginary line that circles Earth halfway between the North Pole and the South Pole (p. 21)

Fahrenheit (adj.) referring to the temperature scale in which the freezing point of water is 32 degrees and the boiling point of water is 212 degrees (p. 15)

heat (n.) energy that comes from the motion of particles that make up matter. Heat flows from an object of higher temperature to an object of lower temperature. (p. 8)

insulator (n.) a material with low conductivity for heat or electricity (p. 19)

joule (n.) the unit used for measuring work or energy in any form. One joule (J) is equal to a force of 1 newton exerted on an object moved a distance of 1 meter—for example, lifting an orange over your head. (p. 9)

matter (n.) any substance that has mass and takes up space (p. 7)

molecule (n.) a group of two or more atoms held together by chemical bonds (p. 6)

radiation (n.) a wave made of oscillating electric and magnetic fields that travels at the speed of light. Radio waves, microwaves, infrared radiation, visible light, ultraviolet light, X-rays, and gamma rays are all types of electromagnetic waves, or radiation. Radio waves have the lowest energy and longest wavelengths. Gamma rays have the highest energy and shortest wavelengths. (p. 22)

specific heat (n.) the amount of heat required to raise the temperature of 1 gram of a substance by 1 degree Celsius. All substances have their own unique specific heat. (p. 24)

temperature (n.) a measure of the average kinetic energy of the molecules in matter (p. 10)

thermometer (n.) a tool that measures temperature (p. 14)

vibrate (v.) to move with a quick back-and-forth motion (p. 7)

Index

air 10, 13, 16, 19, 20, 21, 22, 29
airtight seal 16
alcohol 14, 17
aluminum 24

boiling point 15

calories 9
Celsius 15, 24, 25
  scale 15
chemical energy 8
circuitry 26
cold 5, 11, 12, 13
conduction 17, 20, 21, 22, 26
conductor 18
convection 20, 21, 22, 26
copper 18, 24, 28
currents 21

dissipate 26

Earth's surface 21
electronics 26, 29
electricity 26, 29
energy 7, 8, 9, 10, 11, 12, 14, 15, 20, 22, 23, 29, 30
  forms of 8, 9, 12
  of motion 11, 14, 30
  radiated 22, 23
  transfers of 20, 22, 23
  waves of 22
equator 21

Fahrenheit 15
  scale 15
fire 5, 23, 25
food 9

freezing point 15
fuel 9

gas 7, 20, 22
  molecules in 7

heat 8, 9, 11, 12, 13, 14, 16, 17, 18, 19, 20, 21, 22, 23, 24, 25, 26, 27, 28, 29, 30
  transfer of 13, 14, 16, 17, 18, 19, 21, 22, 26, 27, 28, 30
heat pipes 26, 27, 28, 29
hot 5, 6, 7, 10, 11, 24, 26

ice cube 12
insulators 19

joules 9, 24

light 22
  ultraviolet 22
  visible 22
  waves 22
liquid 7, 20, 22, 26, 29, 30
  molecules in 7

matter 7, 17, 20
  in motion 7
mechanical engineer 27
melt 5
metal 18, 28
microwave 6, 22, 23
molecules 6, 7, 8, 10, 11, 12, 14, 15, 16, 17, 20, 23, 30
  air 10, 16, 20
  alcohol 14, 15
  motion of 7, 11, 12
  water 7, 14, 23

motion 6, 7, 11, 12, 20, 30
  average energy of 11, 14
  matter in 7
  of molecules 7, 11, 12

Ochterbeck, Jay 27, 28, 29

plastic foam 19, 25

radiation 22, 23, 26

sand 24
satellites 26, 27 29
solid 7
  molecules in 7
specific heat 24, 25
speed 11, 12, 14
steam 6
steel 18
Sun 9, 11, 21, 22, 23, 24
sunlight 21, 22
sunshine 22

temperature 10, 11, 12, 13, 14, 15, 18, 24, 25
  scales 15
thermometer 14, 15, 17

ultraviolet light 22

vapor 26
vibrate 7, 16
visible light 22

warm 5, 8
water 7, 11, 12, 14, 15, 17, 24, 30
weather 21
winds 21

About the Author Glen Phelan's fascination with science was sparked when he was a teenager by the lunar missions of the Apollo Program. He shares his fascination through teaching and writing. Learn more at www.sallyridescience.com.

Physical Science
Other Titles

Elements and Compounds

Energy Basics

Energy Transformations

Forces

Gravity

Light

Motion

Physical Properties of Matter

Solids, Liquids, and Gases

Sound

Units of Measurement

www.SallyRideScience.com

ISBN 978-1-933798-54-7

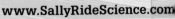

Sally Ride
Science